Spring Meeting Poster Session and Networking Reception 2021

Held at the 2021 AIChE Spring Meeting and 17th Global Congress on Process Safety

Online
18 - 23 April 2021

ISBN: 978-1-7138-5344-2

Printed from e-media with permission by:

Curran Associates, Inc.
57 Morehouse Lane
Red Hook, NY 12571

Printed with permission by Curran Associates, Inc. (2022)

For permission requests, please contact AIChE
at the address below.

AIChE
120 Wall Street, FL 23
New York, NY 10005-4020

Phone: (800) 242-4363
Fax: (203) 775-5177

www.aiche.org

Additional copies of this publication are available from:

Curran Associates, Inc.
57 Morehouse Lane
Red Hook, NY 12571 USA
Phone: 845-758-0400
Fax: 845-758-2633
Email: curran@proceedings.com
Web: www.proceedings.com

TABLE OF CONTENTS

618663 Crude Unit Heat Exchangers and Fouling: Engineered Solutions

Abstract:

Crude Unit Heat Exchangers and Fouling:

Engineered Solutions

Andrew W. Sloley

All heat exchangers in refinery crude distillation units foul. Fouling on the crude oil side of the preheat train imposes significant capital, maintenance, and operating costs on the crude unit. The conventional approach has been to use fouling factors that account for enough fouling for the unit to have a reasonable duration between exchanger cleaning. Fouling factor use suffers from:

- Difficulty in execution – many fouling factors are either too high or too low for specific services. This leads to higher operating costs of over-use of capital.
- Lack of explicit understanding of extra capital involved.
- Over-sized exchangers with low fluid velocities – increasing fouling rates and defeating the purpose of fouling factor allowances.

Enough fouling will shut crude units down, increasing maintenance costs and opportunity costs of lost production.

Of course, the best fouling is the fouling the plant doesn't have. Engineered approaches to decreasing fouling in crude unit exchangers include:

- Valves and piping to allow for cleaning during operation.
- Optimized baffle-cut and baffle-spacing configurations coupled with high shear-stress designs (the 'no fouling' approach).
- Alternate shell-and-tube exchanger geometries
    - Helical baffle
    - Rod baffle
    - EMbaffle®
    - Twisted Tube®
- Surface treatments (polishing)
- Tube inserts
- Non-shell-and-tube exchangers
    - Spiral plate
    - Welded plate

All these options are reviewed and best practices recommendations for crude unit exchangers are made.

618665 Crude Oil Distillation: Unit Objectives

Abstract:

Crude Oil Distillation: Unit Objectives

Andrew W. Sloley

Crude oil distillation is the first unit in the typical oil refinery. It handles the largest volume of material and prepares feed for downstream units. A crude distillation unit is introduced. Major topics covered include:

- Process objectives
- Feed and products
- Performance criteria
- Processing sequence and steps
- Major equipment

618882 Experimental Extraction of Lignocellulosic Biomass Morphological Characteristics for Realistic Representation in Superquadratic Discrete Element Computational Models

Abstract:

The morphological characteristics of lignocellulosic biomass particles, e.g., wood, agricultural residues, and grasses, play an important role in the accuracy of discrete element simulations of the flow behavior of these materials. However, the estimation of morphological parameters is often challenging due to the inherent variability of the size, shape, and distribution of these biological materials. In this study, we propose a rational approach to determine important morphological parameters as inputs in discrete element modeling (DEM) and examine the validity of our proposed approach using an angle of repose case study.

We selected switchgrass ground and sieved through 40 mesh (425 μm) screen as the lignocellulosic biomass sample in this study. We determined the shape parameters and size distributions of the switchgrass sample using a dynamic particle image analysis process. We then qualitatively and quantitively analyzed the switchgrass surface roughness characteristics by digital scanning, scanning electron microscopy (SEM), and atomic force microscopy (AFM) techniques. We generated polydisperse, superquadratic particle assemblies in the open-source DEM package LIGGGHTS using the extracted morphological parameters in the previous step. Furthermore, we implemented the angle of repose study using the hollow cylinder method in the DEM environment and designed validation experiments with a geometrically identical angle of repose test system. We measured the angle of repose and the associated coefficient of friction of the resulting pile for the experiments and simulations and compared our results for validation.

619325 Choosing the Right Metallurgy in the Oil Refining and Chemical Industry

Abstract:

The metallurgy selection in the process industry is very complex and function of fluid service, weather conditions, and process life expectancy. This paper discusses a general class of metals or material divided into ferrous and non-ferrous based alloys. General types of corrosion mechanisms are explained. A typical refinery material selection based on high temperature sulfidation attack, naphthenic acid attack, and hydrogen embrittlement is discussed. Typical cost of different metals is compared. Specialized material selection such as low temperature operation, oxygen service, acids etc.to avoid catastrophic failures is explained. A common metallurgy used in water and refining industries are discussed. A role of a design and material engineers is compared.

619117 Multifaceted Biochar: Biofuel Source, Graphene Precursor and Methane-to-Hydrogen Catalyst

Abstract:

Abstract

Multifaceted Biochar: biofuel source, graphene precursor and methane-to-hydrogen catalyst

Presenting author: Rahul Kundu, email: rak628@utulsa.edu
Advisor: Dr. Hema Ramsurn
Russell School of Chemical Engineering, The University of Tulsa, OK, 74104

Overview

Fossil fuels have a finite reserve and if other carbon sources are not explored, our future global energy demands will not be met. Biomass represents a viable solution due to its abundance and accessibility. Although various techniques, such as pyrolysis and hydrothermal liquefaction and gasification, have been explored to convert biomass to biofuels, there are various challenges associated with these methods. One of the main problems is the formation of biochar and tar as byproducts which reduce the overall conversion of biomass to value-added products. To make the bio-refinery concept more sustainable, any byproduct like char must be converted to valuable products such as biofuel and/or other carbon-based materials. Therefore, my research is focused on the synthesis of value-added products from biochar obtained by hydrothermal carbonization of biomass model compounds (cellulose and lignin). The main advantage of studying biomass model compounds is to better understand their behavior separately as cellulose and lignin are structurally different and consequently translate that to biomass. In my research, I prepared biochar by hydrothermal carbonization in a batch reactor (220-300°C and 3.5-10.3 MPa). Using this biochar as a raw material, I have worked on synthesizing three different products: (a) biofuel, (b) carbon-based nanomaterial (graphene) on metals, and (c) carbon encapsulated iron nanomaterials for methane to hydrogen conversion. During my research I have designed my experimental set-up and gained experience in various analytical instruments including GC-MS, FTIR, Raman Spectroscopy, BET surface area analysis, TEM, XPS and SEM.

Research Experience

For the first part of my research, I used biochar as source to produce biofuel using tetralin as a hydrogen donor solvent at 400°C. I observed that the conversion of cellulose biochar was higher than lignin biochar because cellulose and lignin are structurally different, and lignin is harder to breakdown. Using analytical techniques such as FTIR and GC-MS, I was able to understand and postulate a mechanism for biochar dissolution in tetralin. I concluded that the dissolution of biochar was higher when the C/O of biochar was low and in the presence of biocrude (produced during hydrothermal carbonization). Further, I studied the kinetics of cellulose biochar dissolution where the products formed at various temperatures during the heating step of the dissolution process were analyzed and the major components in the products such as naphthalene (a byproduct from tetralin), levulinic acid (obtained from cellulose) etc. were quantified using GC-MS to understand the role of tetralin during dissolution. I observed that biochar dissolution is a two-step process in the heating stage where first the unconverted biomass in the biochar decomposed followed by the dissolution of biochar. Based on the experimental data, I proposed a new temperature dependent kinetic model where the activation energies were evaluated by non-linear regression. Since, all the conversion occurred during the heating stage of dissolution reaction, the activation energies were evaluated in the heating stage where the rate constant was function of temperature according to Arrhenius equation.
In recent years, graphene, a carbon-based nanomaterial has gained popularity among researchers. Graphene has been called a "wonder material" due to its unique properties. It was discovered by Dr. Novoselov and Dr. Geim in 2004 who won the Nobel prize in 2010. Graphene is a single layer of carbon (sp2) which could be obtained by exfoliation of graphite. It has the potential to be used in biosensors to detect disease like cancer, in microprocessors, in batteries etc. To produce high purity graphene, chemical vapor deposition (CVD) has been used to produce flakes (up to 100 μm) of graphene on copper. Although CVD produces high purity graphene, this method requires greenhouse gas like methane as a source of carbon along with H_2 (flammable) at very high temperature. Also, in most cases CVD requires vacuum in the reactor. These constraints make the process expensive and dangerous. Therefore, to mitigate these issues, I worked on developing a process to use biochar as a source of carbon to produce few-layer graphene on metal foils at atmospheric pressure which does not involve the use of hydrocarbon gas and hydrogen. This makes the process simpler, safer, environment friendly and most importantly less expensive and scalable. While working on this project, I have learned how to use various analytical techniques to characterize a nanomaterial such as scanning electron microscopy (SEM with Energy Dispersive X-Ray Spectroscopy), X-Ray photoelectron spectroscopy (XPS), Raman spectroscopy and X-Ray Diffraction (XRD). The Raman spectroscopy analysis revealed that the presence of the few-layer graphene on all the foils where the 2D/G ratio was less than 1. XPS showed over 80% purity (sp^2 carbon) on the surface of the foil. This innovative process resulted in a patent for which I am one of the inventors (patent's PCT publication number: WO2020/086841). I helped in tailoring our scientific invention into legal terms by working together with my advisor and our legal and IP team. This innovation led me to participate in the i2e Love's Entrepreneur's Cup Business Competition where I teamed up with 2 students from University of Tulsa business school and competed against 30+ teams. Our team won the interview round and was among the 6 selected teams which qualified for the finals. In this competition, we formulated a business plan to produce graphene using our new process mentioned in large scale.
Industrial and technological developments have resulted in rapid increase in the rate of global energy consumption with the energy demand expected to increase to 33% by 2035. Fossil fuel resources are the dominant source of energy and during 2018, the oil demand grew by 1.5% while that of natural gas grew by 5.3% which is the fastest rate of growth since 1984. Due to the increased use of fossil

fuels, the amount of carbon emissions has grown by 2%, the highest in the past seven years. This also led to an increase in emission of greenhouse gases (GHGs) like CO_2, NO_x and SO_x which contribute to climate change. Unfortunately, with the current technologies, an efficient and economical route for utilizing renewables including solar, bio and wind as a source for large scale commercial energy production is not yet available. This gives significant scope for hydrogen, a clean source of fuel to be the future source of energy. Combustion of hydrogen produces only water and advances in hydrogen fuel cell technology can significantly decrease the GHG emissions. Compared to liquid hydrocarbons (13.1 KWh/kg), hydrogen (39.4 KWh/kg) can produce three times higher chemical energy per unit mass. The thermal decomposition of methane is the only single step process which converts methane to hydrogen without producing CO_x. Therefore, I am currently working on synthesizing biochar-based carbon encapsulated nano sized metal (Fe) catalyst from biochar to produce hydrogen from methane in a lab scale continuous flow reactor system. Characterization techniques such as XRD (to study the crystalline structure of the catalyst), BET surface area (to determine the surface area per gram using N_2), SEM, HRTEM (high resolution transmission electron microscopy) etc. are being used to study the synthesized catalysts. The catalyst is then used in the conversion of methane to hydrogen in a continuous flow reactor.

Teaching and Mentoring Experience

Apart from my research experience, I was also a teaching assistant from August 2017 through May 2020. My responsibilities included managing undergraduate chemical engineering laboratory and grading homework and quizzes (fluid mechanics, heat transfer and thermodynamics). In the undergraduate lab, I assisted students in performing laboratory experiments related to enzyme kinetics, composition analysis,
bomb calorimeter, distillation, packed tower, viscometer, and pump performance. I also had the opportunity to mentor undergraduate and high school researchers.

Scholarly Achievements

- Peer Reviewed Journal articles: 2 published (first authored)
- Patent: 1
- Conference Presentations: 7
- Bellwether Fellowship, University of Tulsa (May 2020- May 2021)
- Grant from Office of Research and Sponsored Programs, University of Tulsa (2019)
- Won the interview round and finalist in the i2e Love's Entrepreneur's Cup Business Competition (April 2019)
- Nova Day of Innovation Award (University of Tulsa, 2019)
- Chapman Scholar Presentation Award (2018)
- Phillips 66 R&D Fellowship (Summer 2017, 2018 and 2019)
- Graduate Research and Teaching Assistantship (2016-Present)
- Grant from Student Research Grant Program, The University of Tulsa (2016 and 2020)

Peer Reviewed Publications and patent

- Kundu, R.; Ramsurn, H. "Non-catalytic Dissolution of Biochar in Hydrogen Donor Solvent" Journal of Analytical and Applied Pyrolysis, 140 (2019): 227-238. DOI: 10.1016/j.jaap.2019.03.020
- Kundu, R.; Ramsurn, H. "Kinetic study of non-catalytic dissolution of cellulose biochar in hydrogen donor solvent" ACS Sustainable Chemistry & Engineering, (2020): DOI: 10.1021/acssuschemeng.0c02907
- **PATENT:** Vacuum-Free, Hydrogen-Free Catalytic Synthesis of Graphene from solid hydrocarbons (PCT publication number: WO2020/086841

(PCT publication number: WO2020/086841

618920 Wax Deposition of a High Wax Content Crude Oil in Cold Finger Apparatus and Its Comparison with Flow Loop

Abstract:

ABSTRACT
Wax deposition endangers the flow assurance and safety of pipeline especially for crude oil with high wax content (high waxy crude oil), which contains up to over 25.0wt% wax content in China's submarine pipeline. Present work aims to study the deposition behaviour of a high waxy crude oil in cold finger and flow loop. The deposit mass and thickness, under various temperature differences (wall temperature 30°C, oil temperature 40-50°C), rotational speeds (100-300rpm/min) and deposition time (1-9h), was determined gravimetrically using cold finger. Deposition results in cold finger were then converted to calculated deposit thickness under flow loop condition with Weispfennig model[1-2] based on Chilton-Colburn analogy, and further compared with the experimental deposit thickness from flow loop. Results indicate that the converted deposit thickness (from cold finger) is comparatively thinner than the measured thickness in flow loop. A hypothesis is proposed that wax crystal precipitation and latent heat release must not be neglected for high waxy crude oil, where Nusselt number considering Stefan number[3] and precipitated crystal amount for phase change material[4-5] is introduced into the Weispfennig model.

Figure 1. Comparison between converted deposit thickness (from cold finger) and thickness measured from flow loop experimental value (A1-A3: Original model result, B1-B3: Model with Nusselt number for phase change material)

References

[1] Weispfennig K. Advancements in Paraffin Testing Methodology[J]. SPE International Symposium on Oilfield Chemistry, 2001,2:13-16.
[2] Couto G H , Chen H , Dellecase E , et al. An Investigation of Two-Phase Oil/Water Paraffin Deposition[J]. SPE Production & Operations, 2008, 23(01):49-55.
[3] Dhaidan N S , Khodadadi J M . Melting and convection of phase change materials in different shape containers: A review[J]. Renewable & Sustainable Energy Reviews, 2015, 43:449-477.
[4] Kong M , Alvarado J L , Terrell W , et al. Performance characteristics of microencapsulated phase change material slurry in a helically coiled tube[J]. International Journal of Heat & Mass Transfer, 2016, 101:901-914.
[5] Andreas H . Numerical study of flow-through wall elements with phase–change materials[J]. Journal of Building Engineering, 2018,20:105-113.

617541 What Is the Real Risk Reduction for 3 Sensors Using the Mid-Value for Control and 2oo3 Voting for Safety?

Abstract:

What happens to the risk when two good ideas are combined?

To reduce spurious trips of SIFs, many plants moved from 1oo1 or 1oo2 voting on the sensors to 2oo3 voting -- a good idea. To improve stability for critical process control loops, many plants went from one or two sensors to three sensors using the mid-value for control (also called median-select) -- also, a good idea.

Without really analyzing it, some facilities combined the two ideas, using the mid-value of three sensors for a control loop and then, using the same three sensors voting 2oo3 for an SIF. The intent of the SIF was to protect against consequences that could be caused by a failure of the control loop. This arrangement violates the fundamental premise of LOPA (layer of protection analysis) and ANSI/ISA 84.00.01 (IEC 61511); an independent protection layer shall be independent of causes of the consequence that the layer protects against.

The new configuration must be analyzed by Fault Tree Analysis (FTA), supplemented by Markov analysis. The FTA considers a failure of each of the three sensors and determines which of the remaining devices in the SIF can detect and prevent the consequence. The PFD (probability of failure on demand) is calculated and compared with the PFD of the total independent 2oo3 sensor SIF.

The paper suggests guidance for appropriate use of the combined configuration and suggests how to approximate the risk reduction

617726 Improving Tool Life Prediction Accuracy with Acoustics and Vibration

Abstract:

Improving Tool Life Prediction Accuracy with Acoustics and Vibration

Reece Goff
Undergraduate Student
Swalm School of Chemical Engineering,
Mississippi State University
rcg268@msstate.edu

Amin Amirlatifi
Assistant professor
Swalm School of Chemical Engineering,
Mississippi State University
amin@che.msstate.edu

ABSTRACT

Forming metals and other relatively rigid materials requires milling tools. Although these machines typically perform the task of shaping materials with relative ease, the tool endures wear through this process, and there is always the looming fact that the tool will eventually fail. Predicting these impending failures is vitally important, as a worn out or failed milling machine will perform its task inaccurately, wasting both valuable time and machined materials. Most of the existing correlations used to estimate the remaining life of the tool use physical parameters (such as cutting speed, depth of cut, workpiece hardness, …) but fail to incorporate emissions. Hypersensitive parameters such as acoustic and vibration emissions allow for an accurate calculation of the remaining tool life in real-time. The present study expands the existing correlations used for 'Tool Life' prediction, such as Taylor's Tool Life Equation (TTLE), by accounting for acoustic and vibration emissions. Due to its relative simplicity, this expansion to TTLE will allow for precise prediction of remaining useful life of the tool, saving both time and money.

- Abstract Text

618859 Blockchain Technology and Agile Principles – Applications in Chemical Process Design

Abstract:

Blockchain Technology and Agile Principles – Applications in Chemical Process Design

Bahman Ghorashi

Professor of Chemical Engineering

Tennessee Tech University

Abstract

Blockchain is defined as "a decentralized, distributed, and oftentimes public, digital ledger consisting of records called blocks that is used to record transactions across many computers so that any involved block cannot be altered retroactively, without the alteration of all subsequent blocks."[1]As such, the attractiveness of blockchain is due to its design which makes it difficult to manipulate and alter the recorded transactions, which in turn creates a high degree of trust among the involved participants. Blockchain applications are many and growing in various sectors of the economy including the chemical industry. Some examples of blockchain applications include: smart contracts, supply chain management applications to identify and track transactions across the supply chain, machine-to-machine communication and various innovative combinations with artificial intelligence (AI) and Internet of Things (IoT). While the list of blockchain applications is growing fast, this technology is still in its infancy and has a tremendous potential to bring positive disruptive change in the way business operations and processes are carried out.

Goldman, et. al.[2]define organizational agility as: "The ability of an organization to respond quickly to unanticipated changes in the marketplace and to anticipate future demands and challenges in its core areas of competencies and react accordingly." Certain characteristics are required in order to develop that ability. Those characteristics are:

- Workforce is multi-skilled
- Robust, reconfigurable teams, equipment and infrastructure
- Partnerships in all stages of manufacturing
- Self-directed work teams
- Management breaks barriers
- Effective integration of technology and information
- Fast product development time (concept-to-cash measured in weeks)
- High quality throughout the lifetime of the product

Additionally, other commitments and requirements need to be in place, such as:

- Building upon core capabilities
- Ability to rapidly apply the organization's capabilities to different products and services
- A sincere customer focus
- A commitment to continuous improvement
- Elimination of non-value adding activities

- High quality products and services
- Adopting advanced technologies
- Employee education and training
- Large investments in R&D

According to Goldman, et al. operating in an agile environment calls for : "Short lifetimes for highly differentiated products, created and produced by an active, highly motivated workforce, organized into companies that have widely distributed authority". As such, an agile organization could be viewed as one that has created the processes, tools, and training to make it possible to respond quickly to customer needs and market changes while still maintaining control over costs, efficiency and quality. Agile organizations strive to anticipate the future demands of customers and invest in people and information systems to respond quickly to market changes.

This study examines the various applications of blockchain technology and agile principles in chemical industry as well as potential ways to incorporate those principles into the design of unit operations and chemical processes leading to design of agile plants.

[1]https://en.wikipedia.org/wiki/Blockchain, Retrieved: July 9,2020.

[2]L. Goldman, R.L. Nagel and K Preiss, *Agile Competitors and Virtual Organizations - Strategies for Enriching the Customer*, Van Nostrand Reinhold, 1995.

619346 Improve Process Safety By Uncovering and Mitigating Relief System Design Deficiencies

Abstract:

Have you ever witnessed/experienced/heard of issues such as relief valve bellows got crushed, pressure trend shows that pressure within equipment exceeded design pressure though there is a relief device set consistent with design pressure, relief valves making huge noise, relief valves cycling, relief valve reseating issue etc? If Yes, you are not alone. Ingenero has invested more than 1.2 million man-hrs enhancing process safety by revalidating relief system design for refiners, petrochemical complexes, polymer processing and specialty chemical facilities across the globe. It's not uncommon to see 20-30% of the total relief devices within a facility with deficiencies or deviations from standards leading to issues described above. If left unattended these can greatly compromise overpressure protection and cause hazardous situation. In order to improve process safety further, it is important to uncover deficiencies in the original design first (or avoid them in new design) which is the focus of this presentation.

618175 "Are WE Really Learning"- EIO(EXCELLENCE IN OPERATION) Achieving in PSM Wide Mlwcar(Management, Leadership, Workforce Commitment, Accountability and Responsibility)

Abstract:

"Are We Really Learning" is an over view of Process Plant disasters and similar kind of recurrences elsewhere exposing our gap in Learning from Past for Securing the Present and ensuring a Sustainable Future.

619276 Wettability Induced Preferential Bubble Nucleation of a Gas from a Multi-Gas Dissolved Liquid System

Abstract:

This research aims at verifying wettability induced preferential pressure-driven bubble nucleation of a gas from a multi-gas dissolved liquid present in a hydrophobic glass vial. The hydrophobic glass surfaces were prepared using (heptadecafluoro-1,1,2,2-tetrahydrodecyl)triethoxysilane (HT). The average air-water contact angle and the average roughness of the surfaces were found to be 114.6° ±0.75° and 2.4-7.4 nm, respectively. Degassed DI water in the glass vial in a pressure cell was saturated with a known composition of carbon dioxide and methane gas mixture at 6000 mbar for 24 hours. The saturation time was estimated using one-dimensional bounded diffusion equation to ensure ≥ 99.99% saturated liquid. To initiate the pressure-driven bubble nucleation process, a 500 mbar step-down pressure was applied to the pressure cell at every 15th minute until the bubble nucleation was observed. Methane and carbon dioxide compositions (vol.%) were measured using a micro gas chromatography (micro GC). A digital microscope was used to observe the bubble nucleation process. All the experiments were performed at room temperature (~22 °C).

The average onset bubble nucleation pressure was 4800±274 mbar. The average feed gas concentrations during saturation were 84.44±0.14% methane and 15.44±0.2% carbon dioxide. The corresponding average gas concentrations at onset pressure for bubble nucleation was found to be 85.24±0.48% and 13.12±0.52%, respectively. The increase in the methane gas concentration may be due to the preferential liberation of methane during the bubble nucleation process.

619111 Experimental Investigation of Fe_3O_4-SiO_2 Magnetic Nanoparticles Performance in Demulsification of Oil/Water Emulsions Using the Glass Micromodel

Abstract:

The use of the Fe_3O_4 and SiO_2 magnetic nanoparticles (MNPs) in produced water treatment and oil separation was enhance the amounts of oil produced from the reservoirs via permitting various oil recovery processes to occur for example interfacial tension (IFT) reduction, formation and stabilization of the emulsions, oil wettability alteration, and reduction in the viscosity. To the best of our knowledge the application of the grafted Fe_3O_4-SiO_2 MNPs has been studied in this work for oil separation from produced water in the reservoir is a novel application. The functionalization of the two nanoparticles (SiO_2 and Fe_3O_4) together develop the oil production process from reservoirs because each one has its certain surface property and different effects on oil separation. The results showed the high efficiency of these nanoparticles in oil removal which give high oil removal rate (> 90%). The removal rate was evaluated by the collaborative effects of the synthesized nanoparticles observed by the various mechanisms of the two nanoparticles in the micromodel. The obtained results for the application of Fe_3O_4-SiO_2 nanoparticles in oil removal revealed that the awareness about the reservoir conditions and the stability of the emulsions formed are important parameters for developing oil recovery process

619603 Insights into the Structure Stability of CH_4-C_2H_6 Hydrates Using Molecular Dynamic Simulation

Abstract:

The structure of CH_4-C_2H_6 hydrates changes with gas composition and pressure, but the mechanisms of the hydrate stability and the effect of these influencing factors are unclear. In this work, the molecular dynamic simulation is employed to explore the structure stability of CH_4-C_2H_6 hydrates in sI and sII structures. The cage occupancy ratio of CH_4 and C_2H_6 molecules in the small and large hydrate cages is determined using the thermodynamic models. The average rotation angle, mean square displacement (MSD), and radial distribution function (RDF) of H_2O molecules are applied to evaluate the stability of hydrate cages. It is revealed that the CH_4-C_2H_6 hydrates are more stable in sI structure when the composition of CH_4 is 20 mol%, whereas the sII structure is more stable with a CH_4 composition of 90 mol%. This is consistent with the theoretical calculation and also the experimental observations. The impact of pressure on the structure change of CH_4-C_2H_6 is detected. The stability of sI CH_4-C_2H_6 hydrate cages with a CH_4 composition of 90 mol% increases with pressure, whereas the sII hydrate cages become less stable when the pressure is elevated. The influence of pressure on the hydrate stability varies with the gas composition. When the CH_4 composition is 20 mol%, the stability of sI structure CH_4-C_2H_6 hydrates hardly changes with pressure, while that of sII hydrates is enhanced with the increasing pressure. The rotational and translational motion of CH_4 and C_2H_6 molecules is greatly influenced by the cage occupancy. The motion of CH_4 molecules is affected by the C_2H_6 molecules in the neighboring hydrate cages, which is reflected by the change in amplitude of fluctuation in the average rotation angle and MSD curves, as well as the RDF of C-C pair between CH_4 and C_2H_6 molecules. It is also observed that the frequency of fluctuations in the average rotation angle and MSD curves of CH_4 molecules is larger than that of C_2H_6 molecules. This is a result of the larger potential energy of CH_4-H_2O pair than that of C_2H_6-H_2O pair due to the different size and polarizability of CH_4 and C_2H_6 molecules.

AUTHOR INDEX